I0882984

THE PLANETS OF OUR SOLAR SYSTEM

SATURN

by Heather C. Morris

BrightPoint Press

San Diego, CA

BrightPoint Press

an imprint of ReferencePoint Press, Inc.
Printed in the United States

For more information, contact:
BrightPoint Press
PO Box 27779
San Diego, CA 92198
www.BrightPointPress.com

LIBRARY OF CONGRESS CATALOGING-IN-PUBLICATION DATA

Name: Morris, Heather C., author.
Title: Saturn / by Heather C. Morris.
Description: San Diego, CA: ReferencePoint Press, 2026 | Series: The planets of our solar system | Audience: Grade 7 to 9 | Includes bibliographical references and index.
Identifiers: ISBN: 9781678211721 (hardcover) | ISBN: 9781678211738 (eBook)
The complete Library of Congress record is available at www.loc.gov.

CONTENTS

AT A GLANCE

- Saturn is the sixth planet from the Sun. It is the second-largest planet in the solar system by diameter.
- Saturn's name comes from the Roman god of agriculture who was also the father of Jupiter.
- Saturn is one of four gas giant planets in the solar system. It does not have a solid surface like Earth. It is made up mostly of hydrogen and helium gases.
- One day only lasts about 11 hours on Saturn. This means the planet spins quickly.
- Saturn is known for its dazzling ring system. The planet is surrounded by seven major rings with several gaps between them.

- Saturn is visible in the night sky. People have observed Saturn since before recorded history.
- The invention of the telescope allowed people to study Saturn in more detail.
- *Pioneer 11*, *Voyager 1*, and *Voyager 2* flew past the planet and took pictures. *Cassini* orbited Saturn for 13 years.
- Saturn has 146 confirmed moons. Titan is Saturn's largest moon.

CASSINI'S GRAND FINALE

In 2017, a space probe circled the planet Saturn. Its name was *Cassini*. It was built to study the ringed planet. *Cassini* had performed its mission well so far. The probe had survived outer space for 20 years. It had taken close-up pictures of Saturn. It had measured the planet's **magnetic field**. And it had studied Saturn's **atmosphere**. *Cassini* had also flown by Saturn's moons. It had sent pictures and data back to Earth.

Cassini **was carried into space by a Titan IVB-Centaur rocket.**

This illustration shows *Cassini* orbiting Saturn from a close distance.

Now the probe was preparing for one final mission. *Cassini* was running low on fuel. The people in charge of the mission worried that *Cassini* might crash into one of Saturn's moons. They did not want the spacecraft to accidentally **contaminate** a moon.

The mission planners came up with a solution. They called it the Grand Finale. *Cassini* would dive into Saturn's atmosphere to destroy itself. The planners ordered *Cassini* to fly closer to Saturn than ever before. The spacecraft looped around the planet twenty-two times. It even dove between Saturn and its rings.

At last, *Cassini* plunged into Saturn's atmosphere. This was the first time scientists had received data from inside the ringed planet. *Cassini* fought to keep its antenna pointed toward Earth. It sent back measurements for as long as it could. Finally, on September 15, 2017, *Cassini* burned up. After orbiting Saturn for 13 years, the spacecraft became part of the planet itself.

Jupiter is the largest planet in the solar system by mass and diameter.

THE RINGED GIANT

Saturn is the second-largest planet in the solar system by diameter. It is surrounded by prominent rings. Saturn can be seen in the night sky without the help of a telescope. People have observed this planet for thousands of years.

Today, scientists know more about Saturn than ever. People have used telescopes to study Saturn for a long time. Spacecraft have visited the ringed planet. They took pictures. They gathered data. But many mysteries about Saturn still remain.

MAJOR FEATURES OF SATURN

Saturn is the sixth planet from the Sun. Its path around the Sun falls between those of Jupiter and Uranus. This path is called an orbit. Saturn's orbit is not a perfect circle. Sometimes the planet is closer to the Sun. Other times it is farther away. On average, Saturn is nine times farther from the Sun than Earth. Because it is so far away from the Sun, it takes 29 Earth years for Saturn to go around the Sun once.

Saturn's moon Titan crosses in front of the gas giant. It takes the moon about 15 days and 22 hours to orbit Saturn.

Saturn is 7,337 miles (11,808 km) wider around its equator than around its poles.

One day on Saturn only lasts about 11 hours. This is how long the planet takes to spin once on its **axis**. Only the planet Jupiter has a shorter day. Saturn's rapid spin causes a force at the **equator**. This force makes the planet bulge in the middle. It flattens the north and south poles.

Saturn is the least dense of all the planets. Even though Saturn is almost as wide as Jupiter, it only has one-third of Jupiter's mass. It is the only planet in the solar system that is less dense than water. If it were placed in a very large pool, Saturn would float. Saturn's low density comes from its atmosphere. The planet is made mostly of hydrogen and helium. These are the two lightest elements in the universe.

Unlike Earth, Saturn does not have a solid surface. This is because Saturn is a gas giant. It is one of the four gas giant planets in the solar system. Jupiter, Uranus, and Neptune are the other three. Spacecraft cannot land directly on Saturn.

SATURN'S ATMOSPHERE

Clouds fill Saturn's atmosphere. They form stripes of yellow, brown, and gray.

Fast winds separate clouds into these stripes. Winds on Saturn can reach speeds of more than 1,000 miles per hour (1,600 kmh). That is four times faster than the strongest hurricanes on Earth.

Strong storms come and go on Saturn. In the 1990s, scientists pointed the Hubble Space Telescope at Saturn. This telescope orbits Earth. It took pictures of a storm forming near the center of Saturn. It was called the Great White Spot. The storm slowly faded after a few years. Similar storms periodically occur in Saturn's atmosphere.

Other storms swirl for much longer. The spacecraft *Voyager 1* and *Voyager 2* took pictures of a storm at Saturn's north pole in 1981. This storm remained visible

Great White Spots tend to recur in the northern half of Saturn during its summer season.

in 2024. It has a unique shape. It looks like a hexagon. Rakesh K. Yadav is a scientist studying this storm. He says, “We see storms on Earth regularly and they are always spiraling, sometimes circular, but never something with hexagon segments.”[1]

Saturn's atmosphere has multiple layers. Pressure builds closer to the middle of the planet. This pressure forces gases to change into liquids. At its center, Saturn has a solid core. This core is probably made of metals like iron and nickel. It might be surrounded by rock.

SATURN VERSUS EARTH

Saturn and Earth are very different. Because Saturn is much farther from the Sun, it is much colder than Earth. The average temperature is about –220°F (–140°C). The average temperature of Earth is about 59°F (15°C).

Saturn's diameter is nine times that of Earth. That size does not include its rings.

If Earth were the size of a nickel, Saturn would be about the size of a volleyball.

Saturn and Earth are alike in some ways. Like Earth, Saturn is tilted. Earth's tilt causes different parts of its surface to receive different amounts of sunlight.

This illustration compares the sizes of Saturn and Earth.

This causes seasons. Saturn has seasons as well. Each season on Saturn lasts about 7 Earth years.

SATURN'S RINGS

Saturn is most famous for its rings. These rings extend in a flat disc that surrounds Saturn. They are made of pieces of rock and ice. There is empty space between these pieces. Scientists believe the rings are left over from shattered space rocks or moons.

Some pieces of Saturn's rings are as small as a grain of sugar. Others are larger than a house. The pieces are less tightly packed at certain distances from Saturn. This produces gaps in the disc of material. People have divided this disc into

RINGS OF SATURN

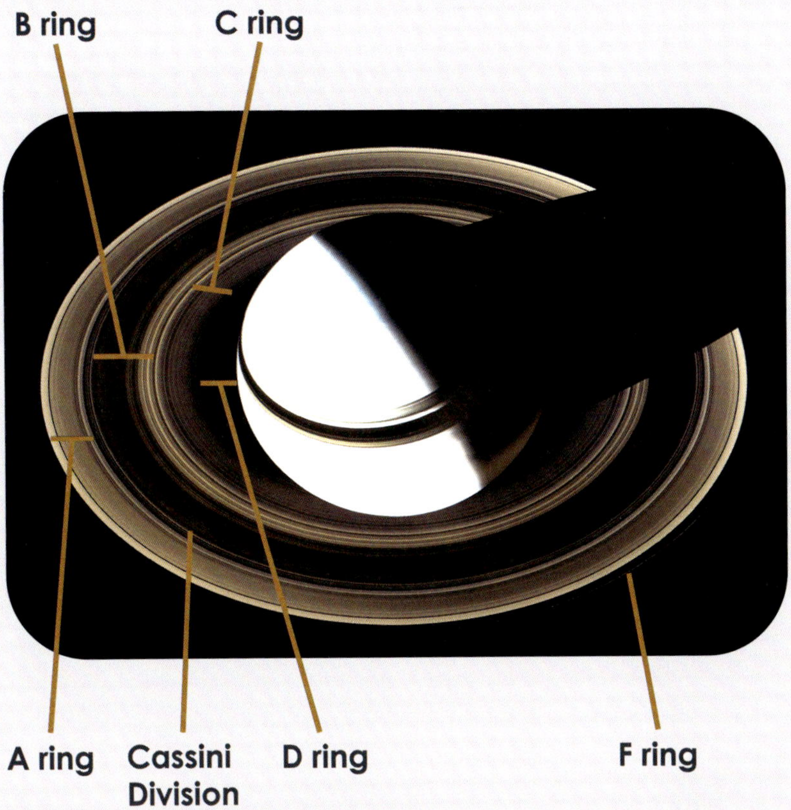

Saturn's rings vary in brightness and distance from the planet. This diagram labels some of Saturn's most prominent rings and ring features.

separate rings. Most of the rings are named after letters. The letters mark the order in which they were discovered. The widest and brightest rings are the A and B rings.

Carolyn Porco speaks about her involvement with a television show about the outer solar system.

Many people think Saturn's rings are beautiful. Carolyn Porco was a leading scientist of the *Cassini* mission. She called Saturn "the most beautiful and the most **photogenic** [planet]."[2] Saturn's rings were discovered in 1610. But people had been observing the planet long before then.

A Small Solar System

Like the Sun, Saturn is made mostly of gas. It is surrounded by many moons and rings. In a way, Saturn resembles a star in a small solar system. Scientists closely study the Saturn system. They watch how gravity causes the moons and rings to interact. This helps them understand the solar system as a whole.

CHAPTER TWO

STUDYING SATURN

People have watched Saturn move across the night sky since before written history. The oldest records of Saturn are from about 700 BCE. The records were made by people tracking the planet. They called it the Star of Ninib. This means Star of the Sun.

About 300 years later, people in ancient Greece also tracked Saturn. They called it a wandering star. The word *planet* comes

One ancient civilization recorded observations of events in the night sky on clay tablets.

from the Greek word for wanderer. The Greeks named this wandering star Kronos. He was their god of farming. The Romans called it Saturn. In their stories, Saturn was the god of farming and the father of Jupiter.

A statue of Galileo Galilei stands at his tomb in Florence, Italy.

SATURN THROUGH A TELESCOPE

In 1608 CE, the first telescopes were built in the Netherlands. People used them to see long distances on Earth. In 1609, astronomer Galileo Galilei pointed his telescope at the night sky. He studied the Moon and the planets. Then he turned his telescope toward Saturn. He thought Saturn's rings were two large moons on either side of the planet. He wrote in a letter that "Saturn is not single but . . . three, which seem to touch each other and never change their relative position."[3]

For more than 40 years, Saturn baffled astronomers. They could not understand its odd shape. In 1655, Christiaan Huygens began to study Saturn. He and his brother made their own telescope lenses.

These lenses allowed Huygens to see the planet more clearly. He looked closely at Saturn. He suggested that Saturn's strange shape was not because of two moons. He thought Saturn was surrounded by a solid ring. Huygens also discovered Saturn's largest moon, Titan.

Twenty years later, astronomer Giovanni Cassini pointed his telescope at Saturn. He could see a gap in Saturn's rings. This gap

The Cassini Division

In 1675, Giovanni Cassini spotted a gap in Saturn's rings. This gap separates the A and B rings. It is called the Cassini Division after the man who first observed it. Hundreds of years later, the *Cassini* spacecraft took pictures of the Cassini Division.

was a clue. It suggested that Saturn's rings were not solid. Later, Cassini discovered four moons orbiting Saturn. He named them Iapetus, Dione, Rhea, and Tethys.

LEARNING MORE ABOUT SATURN

People continued to study Saturn. In 1789, astronomer William Herschel discovered two more moons orbiting Saturn. He named them Mimas and Enceladus. That brought the total number of Saturn's moons to seven. Herschel also saw the gap in Saturn's rings. He suggested that Saturn had two rings that were solid. In 1790, he calculated that Saturn's rings spin around the planet once about every 11 hours. He was the first to notice that Saturn is flatter at its poles.

This illustration displays how grainy material makes up Saturn's rings.

In the 1800s, telescopes became more powerful. They helped astronomers find dark bands in Saturn's rings. These bands turned out to be more gaps. In 1848, a team of astronomers from England discovered another moon. They called it Hyperion.

Saturn's rings continued to puzzle people. No one could figure out what the rings were made of. By 1856, scientists had three ideas. Saturn's rings could be solid. They could be fluid. Or they could be made of tiny pieces of rock.

That year, Scottish physicist James Maxwell used math to prove that Saturn's rings were not solid. His calculations showed that gravity would have torn apart solid rings. He wrote, "The only system of rings which can exist is one composed of an indefinite number of unconnected **particles**."[4] This was the first step in understanding that Saturn's rings are made of rock and dust.

The first photo of Saturn was taken in 1883. Fifteen years later, astronomer William Pickering found another moon orbiting Saturn. He named this moon Phoebe.

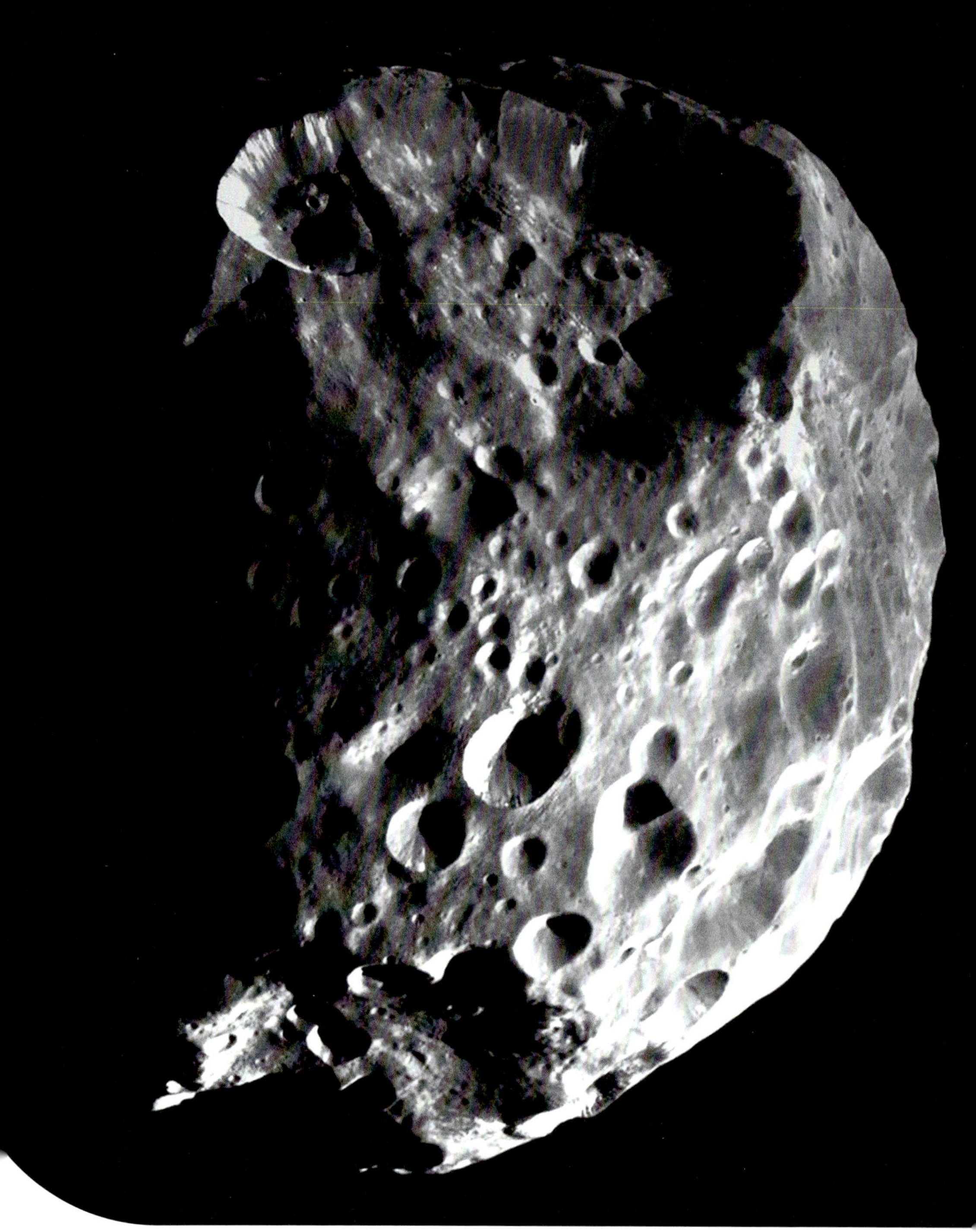

Phoebe's orbit is said to be retrograde. This means it moves in the opposite direction of the solar system's primary direction of motion.

He noticed that Phoebe did not orbit Saturn like the other moons. It orbited in the opposite direction.

People continued to study Saturn in the 1900s. They measured what they could using telescopes. They saw white spots come and go in the planet's clouds. They also discovered more rings. But people wanted to see the planet up close. They wanted to take pictures of the rings and moons. To do this, they would have to send spacecraft to the planet.

CHAPTER THREE

SATURN UP CLOSE

The US government created the National Aeronautics and Space Administration (NASA) in 1958. Its mission is to explore and study space. NASA launches spacecraft from Earth. Some of these spacecraft stay near Earth. Others travel to different planets. And a few have even left the solar system entirely. In 1973, NASA sent the first spacecraft to Saturn. It was called *Pioneer 11*.

This illustration depicts one of the nearly identical *Pioneer 10* or *Pioneer 11* spacecraft sailing through space.

Pioneer 11 weighed 570 pounds (260 kg). It had cameras and other instruments. It also had a radio transmitter. This device allowed it to communicate with people on Earth. *Pioneer 11* launched from Cape Canaveral, Florida. It flew past Jupiter in 1974. Five years later, *Pioneer 11* finally reached Saturn.

Pioneer 11 took pictures and measurements of Saturn from July to October 1979. The spacecraft made many discoveries. It found a new ring. It was narrow. Scientists named it the F ring. *Pioneer 11* also discovered two small moons.

After it flew past Saturn, *Pioneer 11* continued on a path out of solar system. It is still traveling at about 71,000 miles per hour (114,000 kmh). NASA last contacted *Pioneer 11* on September 30, 1995. The spacecraft will shoot past the star Lambda Aquilae in about 4 million years.

THE *VOYAGER* MISSIONS

In 1964, NASA engineer Gary Flandro had an idea. Jupiter, Saturn, Uranus, and

Communicating at the Speed of Light

Distant spacecraft must be able to talk to Earth for people to learn anything from them. The spacecraft send and receive messages using radio waves. These waves travel at the speed of light. This speed is about 671 million miles per hour (1 billion kmh). Messages from Earth can take up to 80 minutes to reach spacecraft near Saturn.

Neptune were lining up to make an amazing mission possible. A single spacecraft could fly past all four planets in 12 years. Normally such a trip would take longer. This arrangement happened only every 175 years. Flandro called his idea the Grand Tour.

Flandro's idea inspired the creation of *Voyager 1* and *Voyager 2*. These spacecraft were identical. They launched a few weeks apart in 1977. *Voyager 2* blasted off first. But *Voyager 1* took a shorter path. It reached Saturn first in November 1980. It discovered five new moons and a new ring. Scientists called it the G ring. Scientists then sent *Voyager 1* to fly by Saturn's moon, Titan. The probe proved that the moon has a thick atmosphere.

Gary Flandro worked at the Jet Propulsion Laboratory. This scientific institution became a part of NASA in 1958.

Voyager 2 reached Saturn about 9 months later. Much of what people know about Saturn today comes from the *Voyager* probes. They measured the length of Saturn's day. They also recorded the planet's temperature. They analyzed clouds and wind speed. And they studied the composition of Saturn's atmosphere.

After leaving Titan, *Voyager 1* was on a path to leave the solar system. In 2012, it became the first spacecraft from Earth to do so. *Voyager 2* took a different path. It visited Uranus and Neptune. This made *Voyager 2* the only spacecraft that has visited all four gas giants. By 2018,

This NASA illustration depicts *Voyager 1* leaving the solar system.

Voyager 2 had also left the solar system. Both spacecraft were expected to continue communicating with Earth until the mid-2030s.

THE *CASSINI* MISSION

After *Voyager 1* and *Voyager 2,* scientists knew much more about the Saturn system. But they still had questions. Instead of flying by, they wanted a spacecraft to orbit the planet. NASA began planning the *Cassini* mission to make this happen.

Cassini carried many scientific instruments. Attached to *Cassini* was a lander named *Huygens*. It was designed to land on the surface of Titan. *Cassini* launched on October 15, 1997. It reached Saturn in 2004. *Huygens* successfully touched down on Titan on January 14, 2005.

Cassini spent 13 years studying Saturn and its moons. It took detailed pictures of Saturn's rings. It measured wind speeds. It also watched a huge storm form on Saturn. The storm wrapped around the planet. It lasted for more than a year. Scientist Carolyn Porco says, "It appears that storms . . . erupt on Saturn the way volcanoes erupt on Earth. Pressure builds and builds until . . . the storm erupts."[5] *Cassini* helped scientists understand even more about the Saturn system.

AFTER *CASSINI*

Cassini's mission ended with its Grand Finale in 2017. But scientists continue to

Saturn's rings display a range of colors due to the materials that make up the rings.

study Saturn. They do this using space telescopes. Space telescopes orbit Earth. They take pictures from above Earth's atmosphere. The atmosphere can blur and distort images taken from Earth's surface. Space telescopes can observe objects without these negative effects.

The Hubble Space Telescope was launched in 1990. Scientists use it to study the outer planets. In 2018, Saturn started changing seasons. Hubble took pictures of Saturn's clouds for 3 years. Each year,

The Hubble Space Telescope takes about 95 minutes to completely circle Earth.

Saturn's equator got brighter. Its winds blew faster. After 2 years, the winds slowed down again.

In 2023, Hubble took pictures of Saturn's rings. Scientists were looking for certain patterns in the rings. *Cassini* and *Voyager* had seen these patterns before. They were called spokes due to the way they seemed to cross multiple rings. Hubble's pictures helped people understand these patterns better.

The James Webb Space Telescope has taken pictures of Saturn as well. This telescope launched in 2021. It uses a different kind of camera from Hubble's. It can capture Saturn's rings in a kind of light that people cannot see. Scientists use these pictures to see Saturn's fainter rings.

CHAPTER FOUR

SATURN'S MANY MOONS

Saturn has more moons than any other planet in the solar system. As of 2023, people have discovered 146 of them. Some are smaller than a sports arena. Some are much larger. Saturn's largest moon is Titan. It is larger than the planet Mercury.

Titan was the first of Saturn's moons to be discovered. It is the only moon in the solar system with a dense atmosphere. Titan's atmosphere is orange and hazy. It is

Titan's thick atmosphere appears as a hazy layer from space.

also very thick. Telescopes on Earth could not see through it to the surface. People wondered what was hidden under Titan's thick clouds.

The *Huygens* lander revealed Titan's solid surface in the pictures it took.

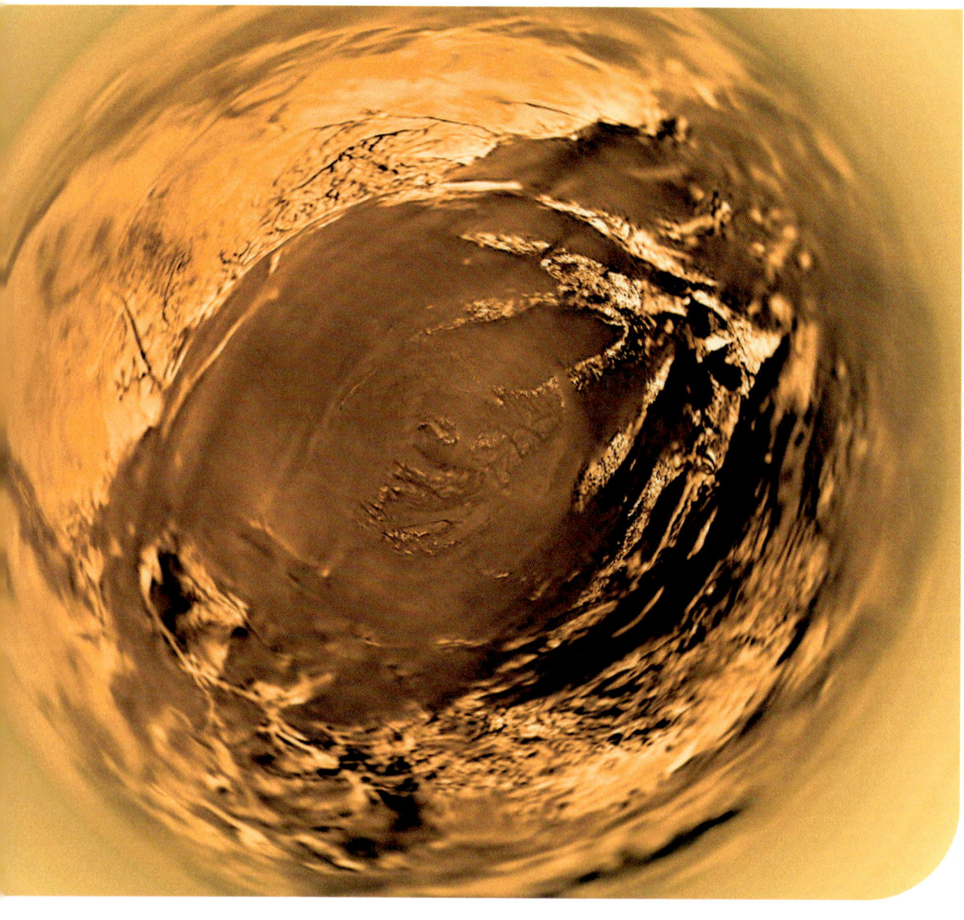

Voyager 1 and *Voyager 2* measured Titan's size, temperature, and air pressure. But neither probe could see through its clouds. Titan's surface was a mystery until 2004. *Cassini* reached Saturn in that year. Its cameras finally peered through Titan's haze. *Cassini* saw seas of chemicals such as methane. It even saw sand dunes.

On December 24, 2004, *Cassini* launched the *Huygens* lander. About 3 weeks later, the lander reached the surface of Titan. No other spacecraft has landed farther from the Sun. *Huygens* sent back the first pictures and data from below Titan's clouds. After a few hours, *Cassini* was out of range. It could no longer receive signals from the lander. *Huygens* still sits silently on Titan's frozen surface.

Dragonfly **will use a drill to take samples of Titan's surface.**

"The *Huygens* descent and landing represented a major breakthrough in our exploration of Titan," says Linda Spilker.[6] She led the science team for *Cassini*.

Cassini and *Huygens* discovered that Titan is the most Earth-like place in the solar system. As with Earth, its atmosphere is mostly nitrogen. Titan also has a liquid cycle. It is like the water cycle on Earth. Liquids rain from its clouds. They fill lakes and **evaporate** into the atmosphere. They form new clouds. But these liquids are not water. They are methane and other chemicals.

Scientists want to know more about Titan. NASA is planning another mission to the moon. It is called *Dragonfly*. It is scheduled to launch in 2028. This flying lander will be as big as a truck. It will have eight helicopter rotors. It will also have cameras and scientific instruments. The team working on *Dragonfly* is designing it to fly by itself. Nicky Fox is a NASA scientist. She says, “Exploring Titan will push the boundaries of what we can do with rotorcraft outside of Earth.”[7]

ENCELADUS

Enceladus is another moon of Saturn. William Herschel discovered this moon on August 28, 1789. It is named after a giant in ancient Greek stories.

Enceladus is a white, frozen world. It has the most **reflective** surface in the solar system. Before *Voyager* and *Cassini*, people did not know much about Enceladus. The *Voyager* spacecraft took pictures of the moon. These pictures showed long, narrow grooves on the surface.

Cassini zoomed past Enceladus many times. In 2005, it discovered ice and gas gushing from grooves at the south pole. Most of this material falls back toward the moon. But some of it flies into space and becomes part of Saturn's E ring.

Scientists refer to the grooves near Enceladus's south pole as tiger stripes.

During later missions, *Cassini* confirmed that Enceladus's geysers also have liquid water. This moon is one of a few places in the solar system with liquid water.

OTHER MOONS

Many other moons orbit Saturn. Hyperion is shaped like a potato. This moon tumbles through space as it orbits Saturn. Its surface is pocked with craters. Some of

Iapetus is Saturn's third-largest moon by diameter.

these craters have bright edges that could be water ice.

Phoebe is another of Saturn's moons. It was the first object that *Cassini* passed when it reached the Saturn system. This moon orbits Saturn in the opposite direction from the other moons. It is one of Saturn's most distant moons. Phoebe has a dark surface that is covered in craters. These facts make scientists think that Phoebe did not form with Saturn. Instead, they think the planet's gravity captured the moon.

Iapetus is another outer moon of Saturn. One side of Iapetus is bright white. The other side is very dark. Scientists think that the two sides **absorb** heat from the Sun differently. The white side reflects heat. The dark side heats up. Ice on that side of the moon melts. This makes the dark material even darker.

Rhea is Saturn's second-largest moon. Its craters might contain water ice. *Cassini* found a thin atmosphere around Rhea. It is made mostly of oxygen and carbon dioxide.

Tethys's enormous crater Odysseus is named after a hero that appears in ancient Greek poems.

Dione is a small moon. It orbits Saturn close to Enceladus. Ice powder from Saturn's E ring constantly coats Dione. This moon is very cold and covered with craters.

Tethys is Saturn's fifth-largest moon. It has a large crater called Odysseus. Tethys's gravity pulls two smaller moons into its orbit. All three moons travel around Saturn together.

Mimas is the smallest of Saturn's major moons. It orbits very close to the planet. Mimas has a gigantic crater that covers one-third of its surface. It is called Herschel. The crater is so large and deep that scientists think the impact almost broke the little moon apart.

Shepherd Moons

Saturn has moons within its rings. Their gravity pulls on the rocks and dust of Saturn's rings. The small moons give the rings clear gaps and edges. Scientists call them shepherd moons. Saturn has five shepherd moons.

GLOSSARY

atmosphere

gases that surround a planet

axis

an imaginary line running through a planet's poles

contaminate

to make something dirty

equator

an imaginary line drawn around a planet that is equally distant from both poles

evaporate

to turn from a liquid into a gas

magnetic field

the area around a magnetic material within which the force of magnetism works

particles

tiny pieces of something

photogenic

fit for being photographed

reflective

capable of throwing back the light that strikes it

SOURCE NOTES

CHAPTER ONE: MAJOR FEATURES OF SATURN

1. Quoted in Juan Siliezar, "A New 3D Model Could Explain the Formation of a Hexagon Storm on Saturn," *The Harvard Gazette*, October 5, 2020. https://news.harvard.edu.

2. Quoted in Susan K. Lewis, "What's Your Favorite Planet?" *PBS*, January 1, 2010. www.pbs.org.

CHAPTER TWO: STUDYING SATURN

3. Quoted in E.A. Partridge and H.C. Whitaker, "Galileo's Work on Saturn's Rings—A Historical Correction," *Popular Astronomy*, April 1896, pp. 408–414.

4. James Clerk Maxwell, "On the Stability of the Motion of Saturn's Rings," *Monthly Notices of the Royal Astronomical Society*, June 1859, pp. 297–304.

CHAPTER THREE: SATURN UP CLOSE

5. Quoted in "Cassini to Saturn: The Journey and the Legacy | Carolyn Porco," *YouTube*, uploaded by *Northwestern CIERA*, October 21, 2022. www.youtube.com.

CHAPTER FOUR: SATURN'S MANY MOONS

6. Quoted in "Huygens: 'Ground Truth' from an Alien Moon," *Jet Propulsion Laboratory*, January 11, 2017. www.jpl.nasa.gov.

7. Quoted in "Dragonfly Mission Confirmed for 2028 Launch to Saturn's Moon Titan," *Johns Hopkins Applied Physics Laboratory*, April 17, 2024. www.jhuapl.edu.

FOR FURTHER RESEARCH

BOOKS

Douglas Hustad, *Fact and Fiction of the Space Age*. Abdo Publishing, 2022.

KS Mitchell, *The Gas Giants: Jupiter, Saturn, Uranus, and Neptune*. BrightPoint Press, 2023.

Arnold Ringstad, *Jupiter*. BrightPoint Press, 2026.

INTERNET SOURCES

"Every Mission to Saturn, Ever," *The Planetary Society*, n.d. www.planetary.org.

"Saturn," *NASA*, December 2024. https://science.nasa.gov.

"Titan: Facts," *NASA*, December 2024. https://science.nasa.gov.

WEBSITES

Dragonfly

https://dragonfly.jhuapl.edu

The Johns Hopkins Applied Physics Lab is in charge of the *Dragonfly* mission to Titan. This website provides information about the mission. It also allows people to explore and move a model of the *Dragonfly* rotorcraft on their computer or cell phone.

Eyes on the Solar System

https://eyes.nasa.gov/apps/solar-system

This official NASA website hosts an interactive model of the solar system with real-time locations of the planets and their moons. Visitors can also learn about space exploration missions on this website.

The Schools' Observatory

www.schoolsobservatory.org

The Schools' Observatory provides educational resources to students who want to learn more about space. It has information about how students can observe objects in space, including Saturn.

INDEX

IMAGE CREDITS

Cover: © NASA
5: © NASA
7: © NASA
8: © NASA
10: © NASA
13: © NASA
14: © NASA
17: © NASA
19: © Mark Garlick/Science Source
21: © NASA
22: © David Buchan/Variety/Penske Media/Getty Images
25: © Photo Researchers/Science History Images/Alamy
26: © jorisvo/Shutterstock Images
30: © NASA
32: © NASA
35: © NASA
39: © Aubrey Gemignani/NASA
40: © NASA
43: © NASA
44: © NASA
47: © NASA
48: © NASA
50: © Steve Gribben/NASA
53: © NASA
54: © NASA
56: © NASA

ABOUT THE AUTHOR

Heather C. Morris writes books for kids who love science and imagination. Before writing full time, she worked as a scientist for a NASA contractor, talking to astronauts on the International Space Station and testing new equipment in extreme environments.